UNDERWATER SMILE SECRETS: AMAZING TEETH FROM THE DEEP

Catherine R. Beebe, RDH, M.Sc.

Underwater Smile Secrets: Amazing Teeth From The Deep
by Catherine R. Beebe, RDH, M.Sc.

Dilettante Living Publications

ISBN: 978-1-951554-08-8

We use our teeth to eat food, but sea creatures use their teeth in many different ways!

Some teeth are shaped to catch slippery prey, while others are designed to cut and tear.

Some teeth are huge and heavy.

Some teeth are tiny and make sand.

Some animals can even use their teeth like antennas, to gather information about the world around them.

Some amazing sea animals have *many, many* teeth.

And some have no teeth at all! Instead they use clever tricks to catch and eat their dinner.

DID YOU KNOW? We have one set of teeth when we are young. These baby teeth fall out as we grow. By the time you are about 12-14 years old, you'll have a full set of adult teeth.

Many underwater creatures have way more teeth than us, always replacing them. Others have teeth that never stop growing.

Let's explore the wild, wacky, and wonderful teeth of the deep!

The Ultimate Tooth Factories

Sharks

Sharks are the tooth making champions of the sea. A shark can have up to 300 teeth in its mouth at one time, lined up in multiple rows!

But here's the interesting part - sharks lose their teeth all the time. They can lose up to 30,000 teeth in their whole lifetime. That's like losing a tooth every single day.

But don't worry about the sharks! New teeth are always growing behind the old ones. When an old tooth falls out, a new one moves forward to take its place.

Different sharks have different shaped teeth for different jobs.

Great White sharks have triangle-shaped teeth like saw blades, perfect for cutting and grabbing slippery prey.

Tiger sharks have curved, notched teeth that can cut through turtle shells!

DID YOU KNOW? Shark teeth are made with a special recipe that makes them super strong. Do you want super strong teeth too? Fluoride toothpaste gives your teeth the ingredients they need to make that same recipe.

DID YOU KNOW? Piranha teeth are so sharp that some South American people have traditionally used them as tools. They would use piranha teeth as tiny knives for cutting or as arrowheads for hunting!

Piranhas

Piranhas have some of the sharpest, most perfectly designed teeth in the river. These famous fish from South America have teeth that work together like a pair of scissors.

Piranhas have triangle-shaped teeth that are lined up in a single row. When they bite down, the top teeth fit exactly between the bottom teeth, creating a scissor-like cutting action that can slice through almost anything!

Each piranha tooth is super sharp and shaped like a tiny triangle with jagged edges, kind of like a mini saw blade. These teeth are so sharp they can cut through tough fish scales, plant stems, and even slice through meat with just one quick bite.

But here's what makes piranha teeth extra special, unlike your scissors they never get dull! Piranhas replace their teeth throughout their lives, so they always have a fresh set of extra sharp choppers ready for action.

Instead of replacing one tooth at a time, piranhas lose and regrow all the teeth on one side of their mouth at once. This way they always have teeth to eat with while the other side is replaced.

Parrotfish

Parrotfish have some of the coolest teeth in the ocean! Their many teeth are stuck together into a strong, hard "beak," just like a parrot's!

They can have up to 1,000 tiny teeth! These teeth are packed tightly like little tiles to form their strong beak.

Like underwater lawnmowers, parrotfish bite off pieces of coral to get at the algae living on it. They don't spit out the hard coral though. A second set of special grinding teeth in their throats crush it into tiny pieces.

Parrotfish teeth never stop growing, so they never get worn down from all that coral crunching. But what do you think they do with all that crushed up coral?

DID YOU KNOW? Parrotfish are beach builders! Many tropical sandy beaches are made mostly from parrotfish sand. After the parrotfish eats the algae, all that ground-up coral comes out as beautiful white sand! One parrotfish can make over 200 pounds of sand every year, and some even make up to 1,000 pounds! That means the sandy beach where you build sandcastles might actually be parrotfish poop!

Manatees

Manatees have some of the coolest teeth in the water. They march! These gentle sea cows have a special tooth system called "marching molars" that works like a slow-moving train inside their mouths.

Manatees don't have any pointy front teeth or sharp canine teeth like meat-eating animals do. They only have flat, grinding teeth called molars because they eat nothing but plants.

These giants spend most of their day munching on sea grass, algae, and water plants that grow on the bottom of rivers and bays.

Manatee teeth are always on the move! New teeth keep growing at the back of their mouths and slowly march forward, pushing the older teeth toward the front.

When the old teeth at the front get worn out from all that chewing, they fall out and get replaced by the newer teeth that have marched up from behind.

DID YOU KNOW? Other animals have moving teeth too! Elephants and kangaroos also get new molars that push forward, but they only get a few sets. Manatees are special because their teeth never stop coming!

Sea Urchins

Did you know that sea urchins have teeth? These spiky little creatures have a super cool feeding system called "Aristotle's Lantern" - named after the famous Greek scientist who first discovered it thousands of years ago.

Sea urchins have five special teeth arranged in a circle inside their mouth. But these aren't ordinary teeth - they're made of hard crystals. This is the same hard stuff that makes up seashells and rocks.

Sea urchins use their special teeth to scrape algae off of rocks and coral, kind of like using a spoon to scrape the last bit of ice cream from a bowl. They can also use them to grind up tough seaweed and small pieces of coral for dinner.

Sea urchins crawl along the ocean floor using their spiny feet, always looking for their next algae snack to scrape up with their amazing teeth.

DID YOU KNOW? Sea urchin teeth are self-sharpening. They chip and break in just the right way so that they always stay sharp. Scientists have studied them to help make better cutting tools and develop materials that don't get dull as quickly.

No Teeth? No Problem.

Baleen Whales

Baleen whales are the giants of the ocean, but they have no teeth at all! Instead, they have hundreds of special plates called baleen that hang from their top jaw like a big curtain.

These baleen plates are made of a tough material called keratin - the same stuff as your fingernails and hair. Instead of biting, baleen whales use their plates like a giant strainer. They open their huge mouths, gulping in large amounts of water filled with tiny sea creatures like krill.

Then they use their big tongue to push water back out through the baleen, which acts as a filter. It traps all the tiny food inside, like a built-in fishing net!

Blue whales, the biggest animals that have ever lived, can have 300 to 400 baleen plates, each up to 3 feet long.

DID YOU KNOW? A blue whale can eat up to 16 tons of krill in a single day. That's like eating over 80,000 sandwiches in one day! They do this all without a single tooth, using only their amazing baleen filter system.

Leatherback Sea Turtles

Leatherback sea turtles, the largest turtles in the world, have no teeth. Instead, they have something amazing: a throat full of spikes!

These sharp spines point backward and line their entire throat. They look like rows of tiny arrows pointing down toward their stomach. This creates a one-way tunnel that traps food and won't let it escape.

Leatherback turtles love to eat jellyfish! Their throat spikes are perfect for jellyfish because they're so slimy and slippery that regular teeth couldn't hold them.

Because they point backwards, the throat spikes act like a jellyfish trap: once a jellyfish goes past them, it can't swim back up and escape! The spikes also help squeeze out extra seawater before the turtle swallows.

DID YOU KNOW? Leatherback sea turtles can eat up to 200 pounds of jellyfish every single day! That's like eating 400 jars of jelly. They are so good at catching jellyfish that they help keep too many jellyfish from filling up the ocean.

Teeth as Super Tools

Narwhals

Narwhals are like the unicorns of the sea. Yet that long, spiral horn sticking out of their head isn't actually a horn at all - it's a giant tooth!

The tusk is an extra-long tooth that grows through the narwhal's upper lip. It never stops growing and can sometimes grow up to 10 feet long. That's taller than a basketball hoop!

Most narwhals with tusks are males, and usually only their left front tooth grows into the long tusk.

Scientists think narwhals use their tusks like an antenna to feel the water around them and find food.

Narwhals don't chew food like we do. Instead, they suck up their dinner like a vacuum cleaner! Sometimes they even tap fish with their tusk to stun them before gobbling them up!

DID YOU KNOW? A narwhal's tusk is super sensitive! It has millions of tiny nerves, which help it feel changes in water temperature, pressure, and even chemicals in the water!

Walruses

Walruses are massive sea animals with giant tusks that they use like built-in tools for almost everything they do.

Walrus tusks are actually enormous canine teeth that never stop growing. Both male and female walruses have these incredible teeth that can grow up to 3 feet long and weigh between 10 to 15 pounds each.

But the coolest part is that walruses use their teeth like tools. They can haul their huge bodies out of the water and onto the ice with them. They also use their tusks to punch breathing holes in the thick ice, fight off polar bears and other predators, and even to show off to other walruses when they hang out in a group.

Male walruses have tusks that are usually longer and straighter than females. They use them in wrestling matches to compete for the best spots on the ice. The walrus with the biggest tusks often wins without even fighting–the other walruses take one look and move out of the way!

DID YOU KNOW? Walruses in zoos get special care for their teeth, just like people go to the dentist! Since their tusks don't have the hard outer layer that human teeth do, zoo workers can put metal caps on the tips to protect them from getting worn down.

DID YOU KNOW? Dolphins don't chew their food like we do. Instead, they use their pointy teeth to grab fish and then swallow them whole!

Dolphins

Dolphins have 80 to 100 cone-shaped teeth that are all exactly the same shape, perfect for catching slippery fish. But dolphin teeth have some pretty cool superpowers too!

Dolphins can send sound waves from their heads and then listen to echoes to help them 'see' where things are around them. This special ability is called echolocation. The way dolphin teeth are spaced apart might work like antennas to help them hear those echoes better.

Dolphins use their teeth in other special ways. Their teeth and gums have really sensitive nerves that help them feel when something touches them or moves in the water around them. Sometimes dolphins even use their teeth for talking to each other - they gently scrape their teeth on other dolphins to show who's the boss!

Dolphins never lose their teeth like we do. They get one set when they're babies and keep them their whole lives! Just like counting rings in a tree to tell how old it is, scientists can count rings inside dolphin teeth to figure out how old a dolphin is.

Masters of the Bite: Specialized Eaters & Hunters

Sperm Whales

Sperm whales are the largest animals on earth that have real teeth. These giants also have one of the strangest tooth setups in the ocean! All of their teeth are in just one place - their bottom jaw.

Sperm whales have between 36 to 60 huge teeth, but only on their lower jaw. The teeth are thick, cone-shaped and made of a hard white material called ivory.

When a sperm whale closes its mouth, the bottom teeth fit perfectly into special sockets on their top jaw, like puzzle pieces locking into place. This unusual setup works perfectly for grabbing their favorite food: giant squid that live deep in the dark ocean.

Sperm whales are great divers. They can hold their breath for over an hour and dive deeper than any other whale to hunt for squid in the dark, deep waters of the ocean. Their powerful teeth help them grab slippery squid tentacles. These tentacles can be as thick as a person's arm.

When we chew our food, our top and bottom teeth work together to mash up our food. Sperm whales solve the problem of no upper teeth by swallowing their dinner whole.

DID YOU KNOW? Each sperm whale tooth can weigh up to 2 pounds - that's as heavy as a large pineapple. Even though they're the biggest toothed animals in the world, sperm whale babies are born without any teeth at all. Their teeth don't start growing until they're about a year old.

DID YOU KNOW? Most fish have throat jaws too, but they can't move them forward like moray eels can. Those throat jaws help fish to crush, grind, and push food down into their stomachs. Only moray eels have special muscles that let them shoot their throat jaws out to grab prey like a secret weapon!

Moray Eels

Moray eels have two sets of jaws. These eels have secret jaws hidden in their throats that help them catch their dinner.

Their main jaws in front have sharp, fang-like teeth that are perfect for grabbing onto slippery fish. These teeth curve backward, making it almost impossible for prey to escape.

But moray eels also have a second set of jaws deep in their throats! When a moray eel catches a fish with its front teeth, those secret throat jaws shoot forward and grab the prey. Then they pull it into the eel's stomach. It's like having a helping hand inside your throat that reaches up to grab your food and pull it down for you!

This double-jaw system makes moray eels incredibly good hunters. They can hide in coral reefs and cracks in rocks, surprising fish that swim by. Their sharp teeth and powerful throat jaws make sure nothing gets away.

Anglerfish

Anglerfish have some of the scariest looking teeth in the deep ocean (and we’ve already seen quite a few scary teeth so far). These mysterious fish live in the deepest, darkest parts of the sea, and their teeth are perfectly designed for this spooky underwater world.

Anglerfish have long, sharp, pointed teeth that curve inward like tiny hooks. What’s different is that their teeth are completely see-through! This makes their teeth almost invisible to prey swimming in the dark water until it's too late.

The curved shape of their teeth acts like a trap. Once a fish swims into an anglerfish's mouth, the inward-curving teeth make it impossible to escape.

Their teeth are so long that anglerfish can't even close their mouths. They swim around with their mouths hanging wide open all the time.

Even though anglerfish aren't very big, they can swallow prey that's twice their own size! Their stretchy stomachs and giant mouths let them eat fish much bigger than themselves. It's like a person being able to swallow a whole turkey!

DID YOU KNOW? Only female anglerfish have these long, see-through teeth! Male anglerfish are much smaller and look completely different. In the deep, dark ocean, the females are the fierce hunters, while the tiny males spend their time looking for a mate.

DID YOU KNOW? Vampire fish aren't really vampires. They got their scary nickname because of those impressive fangs! In the wild Amazon River, these fanged fish are some of the most feared predators - even piranha fish stay away from them.

Vampire Fish

Vampire fish are some seriously scary-looking fish that live in the Amazon River in South America. They have fangs so big and sharp that they look like they belong on a vampire from a spooky movie.

Vampire fish are also called payara. They have two large fangs that stick straight up from their bottom jaw like long sharp spikes. These fangs can grow up to 6 inches long and can poke out of the fish's mouth even when it's closed!

These fierce river hunters use their extra long teeth to catch other fish. When a payara spots its prey, it speeds through the water like an underwater rocket and grabs the fish with its sharp fangs.

The curved shape of these teeth makes it impossible for prey to wiggle free once they've been caught. Once they've speared their dinner, they can swallow their prey whole.

Sheepshead Fish

No, they're not soft and fluffy, but sheepshead fish have the most surprising teeth in the ocean. When you see a sheepshead fish open its mouth, you might think you're looking at a person smiling back at you. That's because sheepshead fish have teeth that look a lot like human teeth!

Just like people, sheepshead fish have different types of teeth for different jobs. They have small, sharp incisors in the front for biting, and big, flat molars in the back for chewing.

When they are first born, they have no teeth. They feed on soft prey at first, but as they get older, their teeth start to grow in and they can eat harder foods.

Sheepshead fish use their teeth to crunch on hard shellfish like oysters, clams, and crabs (something I don't recommend you use your human teeth for). They use their teeth like underwater nutcrackers, cracking shells to get to the tasty meat inside.

Sheepshead fish live near docks, piers, and rocky areas where lots of shellfish grow.

DID YOU KNOW? Sheepshead fish teeth are so much like human teeth that people often do double-takes when they see photos of them! It looks like the fish is smiling back at them.

Sea Otters

Sea otters are adorable sea animals with incredibly strong teeth, perfect for crushing shells.

They have 32 flat, rounded teeth that are mostly molars. Their molars are strong enough to smash clam shells, crab shells, and spiny sea urchins.

This is different from river otters, who eat fish and have sharp front teeth for cutting their food.

Sea otters have an incredibly powerful bite, chomping down with much more force than you would expect for such cute little creatures. Plus, their tooth enamel (the hard outer coating of teeth) is more than twice as strong as human teeth.

Even so, all that crushing of hard food can chip and wear down their teeth, so sea otters carry rocks in a pouch under their arm. This way they can use the rocks like hammers to crack open tough shells and preserve their teeth. Sea otters aren't just cute, they're clever too!

DID YOU KNOW? Sea otter teeth can turn bright purple. When they eat lots of sea urchins, the purple color from the urchins can stain their teeth. Scientists can tell what a sea otter has been eating just by looking at their teeth!

Limpets

Limpets are tiny, but they have the toughest teeth in the entire animal kingdom. These small, cone-shaped sea creatures stick firmly to rocks.

Their teeth are up to 16 times stronger than human tooth enamel! They are even harder than many metals we use to build things.

That's because limpet teeth are made of a super-strong material called goethite. Scientists say it's the toughest natural material ever found!

Limpets slowly crawl across rocky surfaces, using their teeth to scrape up thin layers of algae and bacteria that grow on the surface of rocks.

DID YOU KNOW? Even though limpets are tiny, about the size of a bottle cap, scientists are studying these tiny limpet teeth to learn how to make stronger materials for cars, airplanes, and buildings!

Adult Human Teeth

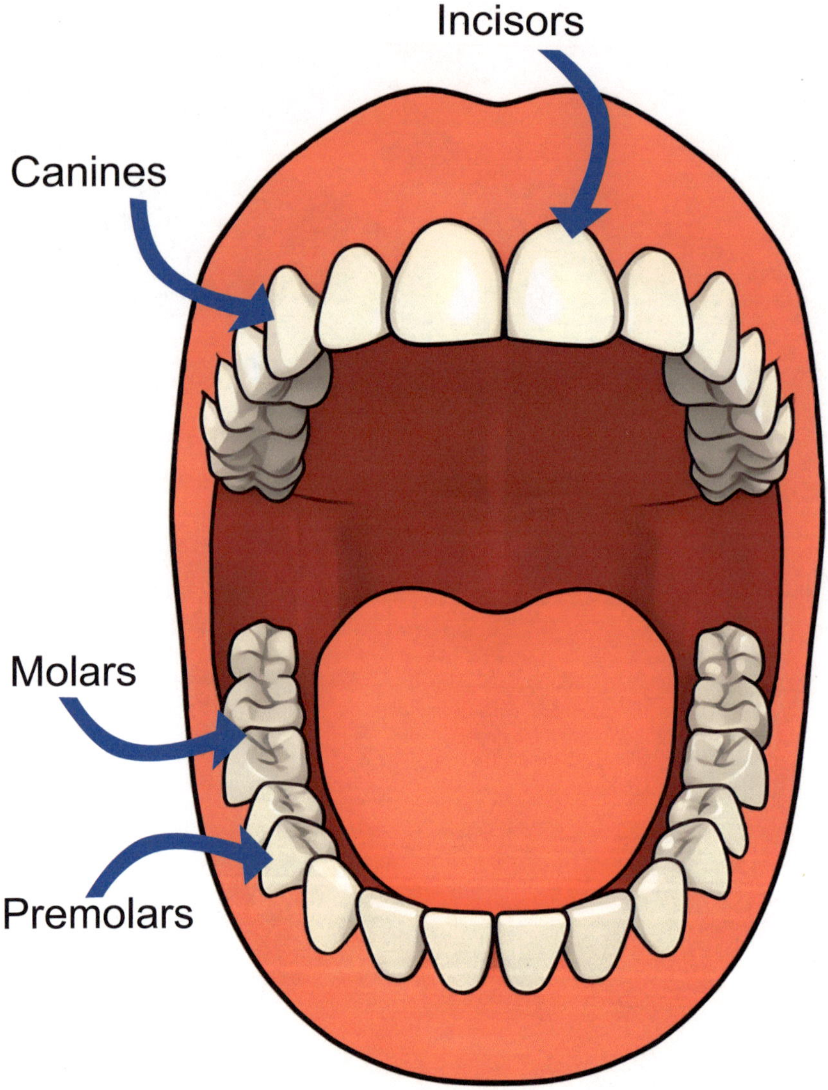

DID YOU KNOW? By the time you are a grown-up, you should have 32 teeth, including four special teeth called wisdom teeth!

Compare Your Teeth!

Look in a mirror at your teeth.

How many teeth do you have?

Young children usually have 20 baby teeth. These will be replaced as you grow older.

Adults usually have 32 teeth.

Are your teeth different shapes?

Your front teeth cut food. These are your front teeth and are called incisors.

Your pointy teeth tear food. These are your canines and are next to the incisors.

Your back teeth grind food. These are called premolars and molars.

Just like fish, your teeth have different jobs!

Primary Human Teeth

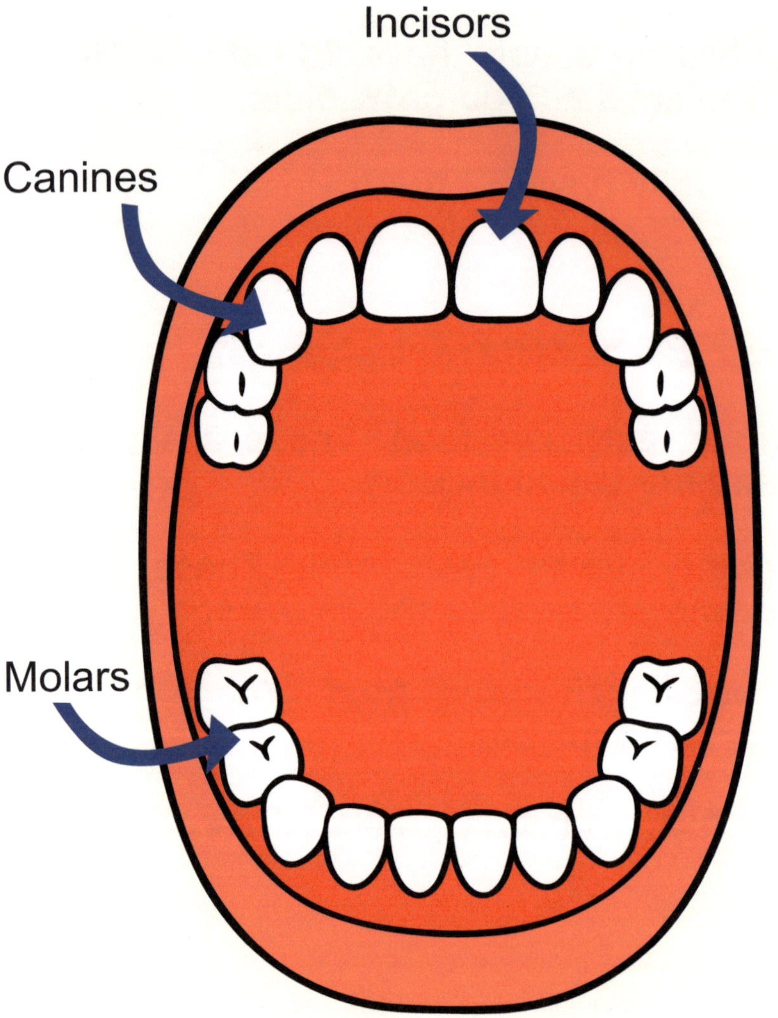

Losing Your Baby Teeth

By the time you were three years old, you probably had all twenty baby teeth, sometimes called primary teeth.

All twenty of these teeth will be replaced by your adult teeth (called permanent teeth) by the time you are 12 to 14 years old.

As your jaw and head grow bigger, there will be room for more molars in the back of your mouth. New adult molars grow in around ages 6, 12, and 18.

Tooth	When It Falls Out
Incisors	5-8 years old
Canines	10-13 years old
Molars	10-13 years old

DID YOU KNOW? Most kids lose their teeth in a pattern. Usually, bottom teeth fall out before the matching top teeth. For example, your bottom middle teeth will probably fall out before your top middle teeth do!

Made in the USA
Monee, IL
05 November 2026

67380131R00029